What Do We Know About the Loch Ness Monster?

by Steve Korté

illustrated by Andrew Thomson

Penguin Workshop

For Marco—SK

For Rhia and Cerys—AT

PENGUIN WORKSHOP
An imprint of Penguin Random House LLC, New York

First published in the United States of America by Penguin Workshop,
an imprint of Penguin Random House LLC, New York, 2022

Visit us online at penguinrandomhouse.com.

Library of Congress Cataloging-in-Publication Data is available.

Printed in the United States of America

ISBN 9780593519202 (paperback)   10 9 8 7 6 5 4 3 2 1 WOR
ISBN 9780593519219 (library binding)   10 9 8 7 6 5 4 3 2 1 WOR

Contents

What Do We Know About
the Loch Ness Monster? 1

Loch Ness 4

Monsters in Scotland 8

Cryptids 22

The 1933 Sightings 32

Nessie Becomes a Celebrity 44

Searching for the Monster 50

What Science Tells Us 69

Is Nessie a Plesiosaur? 79

If Not Nessie 90

The Loch Ness Monster Today 97

Timelines 106

Bibliography 108

Bob Rines

What Do We Know About the Loch Ness Monster?

On June 23, 1972, an American man named Bob Rines was visiting a friend who lived near a large lake in the Scottish Highlands. The name of the lake was Loch Ness. During his trip, Rines saw something in the lake that would change his life forever.

"There in the middle of the bay, we saw a giant hump like the back of an elephant that was somewhat triangular in shape but four to five feet out of the water at the apex," said Rines. "The fear in the back of my neck crawled up, and I shivered. There I could see the gray texture of this animal, like an elephant, a cross between an elephant and a whalelike texture. It moved off against the wind currents, entered the bay, turned around, and very politely came back in front of us to continue viewing us. And then in front of us—plop!—it submerged. We were just speechless!"

Bob Rines was not the first to tell the story of a strange and remarkable sighting on the loch. Over the years, there have been many reports of an unknown animal living deep within the waters of Loch Ness. Although no one is quite sure what it is, the creature is known as the Loch Ness Monster.

CHAPTER 1
Loch Ness

Loch Ness is a lake in Northern Scotland. *Loch* is a word in Scotland's Gaelic language for the English "lake." Loch Ness is a very long, deep lake. It is twenty-three miles long, about a mile wide, and almost nine hundred feet deep. There is more water in Loch Ness than in all the lakes in England and Wales put together. It's big enough to hold every person on Earth more than ten times over!

Tall mountains surround Loch Ness, and there are eight small towns located around the loch. The largest town is Fort Augustus, with a population of only 646 people. Tourism, sightseeing by boat, and fishing are the main industries of the people who live near the loch.

Loch Ness is connected to the North Sea by the River Ness and a man-made waterway called the Caledonian Canal. The cold water that flows into the loch contains particles of peat, which are bits of decayed vegetation. The peat turns the water

Peat

of Loch Ness dark brown. The peat also blocks sunlight from penetrating more than about thirteen feet, making it nearly impossible to see anything below the water's surface.

Many believe that deep, dark Loch Ness may hold more than one mystery. There are ancient legends and mysterious events connected to the loch. People have drowned in its waters and then completely disappeared. Their bodies have never been recovered. Locals who live near Loch Ness wonder if the unfortunate people who drowned could have been devoured by a large animal hiding in the lake's murky depths!

What Lives in the Loch?

An estimated twenty-seven tons of fish live in Loch Ness, including salmon, sea trout, char, and eels. (More than enough fish to feed any large, aquatic creature that might be hiding in Loch Ness . . . if such a creature exists!)

There are also amphibians living at the edge of the loch, including frogs and toads. There have been sightings of seals swimming in the loch, and several land creatures—including pigs, deer, and humans—occasionally take a dip in Loch Ness, too.

CHAPTER 2
Monsters in Scotland

Scotland is a land of secrets and legends. Hidden within the dark depths of Loch Ness lies a mystery unsolved for centuries. It is the Loch Ness Monster, or "Nessie," as the creature is sometimes called. Nessie is part of a very long

"Nessie"

Scottish tradition. For many years, there have been tales of monsters in Scotland. Some of these creatures lurk within the country's lakes and rivers.

We know about the history of monsters in Scotland because ancient

pictures of these strange beasts have been discovered. The drawings were made by the Picts, people who lived in Scotland from around AD 100 to 840. The Picts carved images of large water beasts into

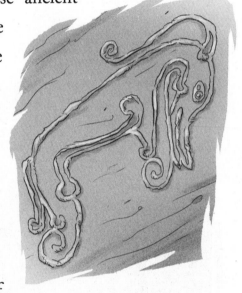

A Pictish beast

stone almost 1,500 years ago.

One of the most fearsome water creatures was the kelpie. According to Scottish legends that come from the time of the Picts, the kelpie was an associate of the devil. It lived underwater and was often described as a cross between a bull and a horse, with two sharp horns on top of its head. It would emerge from the water when it was hungry. The kelpie would magically transform itself into a beautiful horse with an elaborate saddle on its back. It would calmly graze by the side of a loch or river, waiting for an unsuspecting person to climb onto its back. The beast would then gallop back into the water, where its victim

Kelpie

would be drowned and eaten by the kelpie. In other versions of the legend, the kelpie would take the appearance of a handsome young man and lure women to their underwater deaths.

Kelpies were said to exist in most bodies of water in Scotland. Since there are more than thirty thousand lochs and rivers in Scotland, that's a lot of kelpies!

There were other beasts in ancient Scottish legends. One was a killer seal that was known as a selkie. It was said that these dangerous monsters could shape-shift and transform from cuddly seals

into humans. In their human form, the selkies would sometimes escort their victims into the water and devour them.

The buarach-bhaoi was a nine-eyed eel that would jump out of the water and wrap its body around a human's ankles. After the buarach-bhaoi dragged its prey into the water, then it would suck its victim's blood.

Buarach-bhaoi

The dreaded boobrie was an oversize bird that ate meat and lurked near lochs in Scotland.

Boobrie

According to one report, the boobrie was "larger than seventeen of the biggest eagles put together." It possessed a razor-sharp beak that it used to rip apart small victims, including lambs, otters, and children who wandered too close to the shore of a loch.

And there is the famous story of Saint Columba. In 565, an Irish monk named Columba was traveling through Scotland when he heard about a dangerous water beast that lived in the

River Ness, which is connected to Loch Ness. It was said that the dragon-like creature had already killed a local resident. Columba decided to travel to the River Ness to offer his help.

Columba arrived at the shore of the River Ness and asked his faithful servant, a man named Lugne Mocumin, to swim across the river and return with a boat. But when Mocumin jumped into the water, he disturbed the monster, which was hiding at the bottom of the river.

Saint Columba (521–597)

Columba was born in what is now the county of Donegal in Ireland. Said to be of royal blood, he was a deeply religious man who became a missionary. After a bloody battle in 561 resulted in the death of three thousand men, Columba fled to the west coast of Scotland, where he established a monastery on the island of Iona. (He had been worried that the angry families of the fallen Irishmen would blame him for the loss of lives in battle.) It was while he was doing missionary work in Scotland that Columba confronted the water beast at the River Ness.

About a century after his death, an abbot at the Iona monastery wrote a biography of Columba and created one of the earliest written accounts of what has come to be known as the Loch Ness Monster.

A statue of Saint Columba can be seen today at Fort Augustus Abbey, at the southern end of Loch Ness.

Suddenly, the beast burst out of the water. Its mouth opened wide, revealing razor-sharp teeth. As it moved toward Mocumin, Columba is said to have cried out, "Go no further! Nor shall you touch the man. Turn back at once!"

The beast was so terrified by Columba's words that it instantly backed away and returned to

the depths of the deep river. Some said that the beast never again harmed a person in the river or loch, thanks to the powerful words spoken by the monk.

The legend of that monster survived for centuries, and new sightings of the creature continued to surface over the years.

In 1879, a group of schoolchildren was playing in a graveyard on the shore of Loch Ness. Suddenly, a strange animal came into view. It had a giant body and a tiny head, and its skin was gray. The astonished children watched as the heavy beast waddled down a hill and into the loch.

In 1919, a teenager named Margaret Cameron was playing on the shore of the loch with her two brothers and younger sister. Margaret heard a sound in the trees on the other side of the bay and saw a large creature move awkwardly down toward the water. "It had a huge body, and its movement as it came out of the trees was like a caterpillar," she later said. The beast had shiny skin that was the color of an elephant, with two short, round feet in front. It twisted its head from side to side and then sank into the water.

Was the monster that confronted Columba fact or legend? And were the creatures seen by the schoolchildren and Margaret Cameron centuries later real? Could such a beast actually exist in the world?

CHAPTER 3
Cryptids

There have long been reports of strange creatures all around the world that are unknown and unclassified by the scientific community. There's even a name for them: They are called cryptids.

A cryptid is an animal that science has been unable to prove actually exists. The study of these creatures is called cryptozoology (say: krip-tow-zoo-AH-luh-jee.) That comes from the Greek word *kryptos*, which means hidden, and the word *zoology*, which is the study of the behavior of animals.

The best way to prove that a cryptid is real is to obtain an actual specimen, such as a hair or skin sample, of the animal—whether it's alive or dead. Over the years, several mysterious cryptids have managed to surprise the world by turning up very much alive.

During the 1800s, there were reports of a ten-foot-long "land crocodile" in the Southeast Asian island country of Indonesia. Few scientists in other countries believed that this dinosaur-like creature could be real. Then, in 1910, live mega-lizards were observed by Europeans on the island of Komodo in Indonesia. Scientists later gave the species the official name of Komodo dragon.

Komodo dragon

Okapi

For years, there were rumors about a long-necked animal with strange, striped legs that was said to live in the Congo River basin in Africa. European explorers first found the creature in 1901, and scientists formally recognized it as a relative of the giraffe and named it the okapi.

There had been many stories about a manlike ape in Africa, but it wasn't until 1902 that European explorers encountered the mountain gorilla.

Mountain gorilla

As new animals continue to be discovered and recognized by science, many people wonder if the deepest and most remote waters on Earth may contain other undiscovered creatures.

In the novel *The Loch*, author Steve Alten suggests that "the animal referred to as Nessie, if it

Steve Alten

exists, is an undiscovered species of sea creature, perhaps even a mutation. Even in this day and age, large, extinct land and water creatures are being discovered all the time, thanks to advances in technology and our ability to gain access to hostile environs." Hostile environs (environments) are places with extreme climates or landscapes, which are often unexplored.

The Loch Ness Monster is unquestionably the most famous water-based cryptid in the world. Equally famous are two hairy land-based cryptids. One is Bigfoot, rumored to live in dark forests in the Pacific Northwest. The other is the yeti, a large animal that is said to stalk the snowy Himalayan mountains.

Yeti

Other well-known cryptids include a monster
squid known as the kraken and a colossal shark
called the megalodon. There are thousands of
eyewitness accounts from all around the world
describing these strange creatures.

Tales of Scotland's monsters have been around since the time of the Picts. In modern times, Scottish newspapers occasionally carried stories about sightings of unusual lake creatures, but few outside of the country paid much attention to them. That changed in 1933, the year that the Loch Ness Monster really started making headlines.

CHAPTER 4
The 1933 Sightings

John and Aldie Mackay lived at the Drumnadrochit Hotel on Loch Ness. On April 14, 1933, they were driving back to their hotel on a

narrow road around the loch when Mrs. Mackay spotted something moving on the surface of the water. Looking closer, she saw that it was an enormous animal with a dark body that resembled a whale's.

"It just rose out of the water's black wake, with the water rolling off it," Aldie later told a reporter. (A wake is a trail left by something moving through water.) The creature surfaced and plunged back into the water until it disappeared into the lake with a giant splash.

A few weeks later, the *Inverness Courier* newspaper published an article about what Mrs. Mackay had seen under the headline "Strange Spectacle on Loch Ness: What Was It?" This story was the first to call the creature in the loch a "monster."

It would not be the last newspaper story in Scotland about the Loch Ness Monster.

What's in a Name?

The Loch Ness Monster is often affectionately referred to as "Nessie." Some claim that the nickname first appeared during the 1933 sightings of the monster.

Long before that, though, the creature was called *an Niseag*, which is a nickname taken from the Gaelic name for the lake, Loch Nis. The word *Niseag* may have inspired the Nessie nickname. Nessie is also a variation of the female name Agnes, which perhaps explains why so many people refer to Nessie as a female.

In 1975, the World Wildlife Fund gave the Loch Ness Monster its own scientific name: *Nessiteras rhombopteryx*.

Three months after Mackay's sighting on the lake, on July 22, 1933, a man from London named George Spicer and his wife were driving next to Loch Ness. Their car almost hit a giant animal as it slithered across the road and into the loch. Mr. Spicer sent a letter to the *Inverness Courier* describing the incident.

George Spicer

"I saw the nearest approach to a dragon or prehistoric animal that I have ever seen in my life," wrote Spicer. "It crossed my road about 50 yards ahead and appeared to be carrying a small lamb or animal of some kind. It seemed to have a long neck, which moved up and down . . . and the body was fairly big, with a high back."

George Spicer's description of the creature's long neck caused some people to wonder if Nessie might be a type of long-extinct animal that had lived in seas during the time of the dinosaurs.

The newspaper story of a dinosaur-like creature crossing a road in Scotland caused a sensation. People began traveling to Loch Ness, hoping to catch a glimpse of the monster.

In December 1933, the first known photograph of Nessie appeared in Glasgow's *Daily Record and Mail* newspaper. The picture was taken by Hugh Gray, who claimed that he had photographed a forty-foot-long creature that he had seen thrashing around in Loch Ness. The photo was so blurry, though, that it was hard to tell exactly what it was. Some claimed that it actually showed a dog—a Labrador retriever—swimming toward the camera with a stick in its mouth. Others thought that it could be a floating log. Gray's photo was of such poor quality that any of those explanations was possible.

Marmaduke Wetherell

Interest in Scotland's Loch Ness Monster was steadily growing. At the end of 1933, the *Daily Mail* newspaper in London hired a big-game hunter named Marmaduke Wetherell to track down the creature. Within days of arriving at Loch Ness, Wetherell announced that he had discovered monster-size footprints on the shore of the loch!

What happened next launched Nessie on the path to international stardom.

The *King Kong* Connection

In 1933, another famous monster made its debut; this one arrived on movie screens. The movie was *King Kong*, and it told the story of a giant ape that lived on a remote jungle island. But Kong is not the only monster in the movie.

During one scene on the island, a long-necked water monster attacks a group of men traveling down the river on a raft, killing several. The animal looks a lot like a prehistoric—and long-extinct—dinosaur.

King Kong was a huge hit around the world. The movie opened in London on April 16, 1933. George Spicer later told a reporter that he had seen the movie in London before he and his wife spotted the Loch Ness Monster nearly six hundred miles away in Scotland. Was it just a coincidence that there were so many Nessie sightings after *King Kong* was released?

CHAPTER 5
Nessie Becomes a Celebrity

The *Daily Mail* announced Marmaduke Wetherell's amazing discovery on December 21, 1933, under the headline "Monster of Loch Ness Is Not Legend But a Fact."

Plaster casts were made of the footprints and sent to London, where they were examined at the Natural History Museum. Only eight months after Aldie Mackay had sighted Nessie, it appeared that the mystery of the monster was about to be solved.

Unfortunately, Wetherell's find turned out to be a fraud. Experts at the museum quickly determined that the footprints matched those of an African hippopotamus! There are, of course, no hippos living anywhere near Loch Ness. Wetherell had tried to fool the world by creating fake oversize footprints. He produced the tracks using the dried foot of a long-dead hippopotamus. Wetherell's moment of fame in Loch Ness ended quickly, and he left the area in disgrace.

Despite the hippo hoax, interest in the Loch Ness Monster continued to build. A new highway was completed around the loch in 1933, and that made it much easier for anyone with a car to explore the once-remote region. There were now more people traveling to Loch Ness and more sightings. Just one year later, the *Daily Mail* published a spectacular photo of Nessie. This would turn out to be the most famous photo ever taken of the Loch Ness Monster.

A doctor from London named Robert Kenneth Wilson claimed to have been driving along Loch Ness on April 19, 1934, when suddenly, he noticed a disturbance in the water. Wilson would later say that he sighted "the head of some strange animal" and that it was "between 150 and 200 yards from the shore." Wilson grabbed his camera and quickly took four pictures of

Dr. Robert Kenneth Wilson

the creature as it rose out of the water and then sank out of view. After photos were developed, one clearly revealed a large, dinosaur-like creature with a long neck and a small head rising out of the loch.

The *Daily Mail* published the photo on April 21, 1934, along with Wilson's account. The black-and-white picture quickly became known as the "surgeon's photograph" because Wilson was a doctor. Experts examined the photo negative and determined that it had not been altered. Because Wilson was a well-respected physician and a military veteran, few believed that he would have made up such a story. For many years, Wilson's photo was accepted as the best piece of evidence that some kind of prehistoric animal was living in Loch Ness.

The Loch Ness Monster became a worldwide sensation. One of the first songs about Nessie, "Boo! Boo! Boo! Here Comes the Loch Ness Monster," was recorded in 1934. Tourists filled the hotels around the loch, and countless stories of the monster appeared in newspapers

One of the original recordings about the Loch Ness Monster from 1934

around the world. A British circus offered a reward of 20,000 pounds (worth over $2 million in US dollars today) for the live capture of Nessie.

A New York zoo offered 5,000 pounds ($505,000 today) if the creature could be brought to America.

The Steel Scaffolding Company in London made headlines when it announced that it had built a cage especially designed to hold and transport Nessie, if she should ever be captured alive. However, few of the cages were sold, and none were ever used in Loch Ness.

The Loch Ness Monster continued to swim free.

CHAPTER 6
Searching for the Monster

After the publication of the surgeon's photograph, Nessie sightings poured in to newspapers from both locals and curious tourists.

On May 26, 1934, a monk named Richard Horan was working in the boathouse of Fort Augustus Abbey on the shore of Loch Ness. He heard a loud splashing noise in the water. Horan looked toward the lake, and he was startled to see a creature, with a seal-like head and a long neck, staring at him. The animal was about thirty yards away. It slowly swam away from him and then sank into the water. Three other people standing in different locations around the loch also saw the same thing. One of the other witnesses reported "a strange object that seemed to shoot out of the calm waters almost opposite the abbey boathouse." This witness said that the beast's head and neck extended about six feet out of the water and its body, which had a large, rounded hump, was about thirty feet long.

In 1938, John MacLean spotted the head and long neck of an unidentifiable creature in Loch Ness. He said the animal's head resembled that of a sheep, without any ears, and it had a pair of narrow eyes at the front. It appeared to be chewing on something, and occasionally it would throw back its head to swallow. After a few minutes, the beast lowered its head into the water and a giant, humped back came into view. MacLean described the animal's skin as smooth and dark.

John MacLean explains his sighting to a reporter

More sightings of Nessie were reported every year after that. Most of the eyewitnesses described a large, dinosaur-like creature that would raise its long neck above the water for a minute or so, and then it would sink back into the dark, inky depths of the loch.

Alex Campbell, a reporter for the *Inverness Courier*, had an unusual experience during the mid-1950s. He was sailing with his dog on the southeast part of Loch Ness when his boat suddenly started to rock violently in the water.

Campbell said it was "as though the creature was directly beneath them." The dog let out a frightened sound and, shivering in fear, quickly hid under Campbell's seat. The boat was knocked into the air and then landed back in the water.

Suddenly, the extreme rocking stopped. Whatever had caused the disturbance disappeared back into the loch.

An English journalist and wildlife researcher named Fredrick Holiday reported seeing the

Loch Ness Monster a total of four times during the 1960s. One of those times, he was gazing out at the loch when he saw something moving just below the surface. Holiday grabbed his binoculars and focused on a creature that rose about three feet out of the loch. Holiday said it was "black and glistening" with a shape that was "thick in the middle and tapered towards the extremities." He estimated that it was about forty-five feet long. The creature then sunk back into the dark water.

More photos of the monster also appeared over the years, along with many disagreements about how authentic they were.

One person who believed that the surgeon's photo of 1934 was real was an aeronautical engineer named Tim Dinsdale. He traveled to the south shore of Loch Ness, hoping to catch a glimpse of the monster. On April 23, 1960, Dinsdale saw an oval-shaped, reddish-brown creature swimming in a zigzag pattern through the water. He pointed a small movie camera toward

Tim Dinsdale

the loch and started filming the animal. He saw "rhythmic bursts of foam" breaking in the water at the side of the creature as it swam. Dinsdale was convinced that these bursts of foam were caused by the monster's flippers as it moved through the water. He estimated that the creature was traveling at a speed of approximately ten miles per hour.

Tim Dinsdale was convinced that he had captured the first-ever footage of Nessie. His four-minute black-and-white film was shown on the British television show *Panorama* on June 13, 1960, where it attracted a lot of attention.

A few years later, photographic experts from England's Royal Air Force examined the film and declared it was probably a living animal and not a submarine or other underwater vessel.

Decades later, the Discovery Channel analyzed the film again and used computers to improve the images. The newly enhanced film revealed a previously unseen shadow located behind the creature. The shape of this shadow appeared to be a large body just below the surface of the water. The new analysis seemed to support Tim Dinsdale's claim that he had filmed a massive water beast.

Interest in Nessie hit an all-time high after the release of the Dinsdale film. A local organization called the Bureau for Investigating the Loch Ness Phenomena was formed in 1961. The group eventually changed its name to the Loch Ness Investigation Bureau (known as LNIB), and over the next decade, it conducted several searches for the Loch Ness Monster.

One of the group's first projects was a two-week investigation in 1962. Twenty-four volunteers were stationed around the loch during the day and evening. At night, they used high-intensity searchlights that had a range of six miles.

On October 18, one of the volunteers spotted a tall, slim object extending out of the water. That same day, seven members of the team saw and filmed a long, dark creature swimming in the loch. The film was analyzed by Britain's Joint Air Reconnaissance Intelligence Centre (JARIC), which estimated that the animal was "six to eight feet long, three foot high and six foot wide." JARIC's role was to study military images and provide information to the British government. They confirmed that it was not a boat or a submarine.

Just a few years later, the LNIB sponsored a mission to try to harpoon Nessie! The LNIB was determined to obtain a tissue sample (cells of bone, skin, or muscle) from Nessie, explained photographer Dick Raynor. "And to this end,

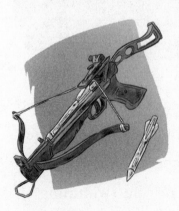

we used to go out in a small rowing boat at night dressed in wetsuits, armed with a crossbow. On the end of this crossbow, there was this tissue-sampling dart, which is hollow, and

Crossbow with dart

the idea was that when Nessie came up, you fired the dart into its hide and pulled on the line to retrieve a tissue sample. Whereafter, you beat a hasty retreat. I did two or three night drifts, but we never got close enough to see anything."

In 1969, the LNIB launched one of the first underwater investigations of Loch Ness. A small yellow submarine called the *Viperfish* was lowered into the loch, with an American named Dan Taylor piloting the craft. Unfortunately, it was impossible for Taylor to see anything

Viperfish

outside the submarine in the dark waters at the lower depths of the loch. The mission failed to locate Nessie.

In 1972, an American scientist named Bob Rines used a motorized underwater camera attached to a powerful light in an effort to photograph the Loch Ness Monster.

Two of his underwater photographs captured what appeared to be a creature with a large, diamond-shaped flipper on the side of its body. The many dark peat particles floating in the water made the pictures incredibly blurry, but experts at the NASA Jet Propulsion Laboratory in California used computers to improve the images. After they did, the flipper could be seen more clearly. It was estimated that it was about six and a half feet long. Because the flipper in the second photo was at a slightly different angle, the experts determined that it had moved. They said it was possible that a large creature with six-and-a-half-foot-long flippers was swimming in Loch Ness. Were these the first underwater photos of Nessie?

Eyewitness reports continued all around Loch Ness. In 1990, Valerie Moffat was driving along the loch when she had a memorable Nessie sighting.

Valerie Moffat

"I came up the hill where we came in sight of the bay, glanced out across it, and saw this large lump, is the best way to describe it," said Moffat. "The nearest I can tell you is it looked like a boat that had turned upside down. . . . About thirty feet in length, and nearly ten feet in height from the water to the top of the back.

It was a bright, sunny day, the water was bright blue, and it really showed up against it. It was a mixture of browns, greens, sludgy sort of colors. I looked at it on and off for a few seconds, because I was driving. Must have seen it three or four times, and the last time I looked, it was gone!"

There have been an estimated 1,400 reported sightings of Nessie to date. Paul Harrison, the author of *The Encyclopaedia of the Loch Ness Monster*, says, "Local people still see something in the loch. Visitors still report sightings of strange humps. Fishermen passing through the loch still see curious wakes. It will continue that way until something is actually pulled from the loch and produces evidence of being the Loch Ness Monster."

CHAPTER 7
What Science Tells Us

Some have wondered if science—specifically scientific studies using sonar and DNA—might help to determine once and for all if Nessie really exists.

Because the water in Loch Ness has been stained by peat particles, it is impossible to see through it. Scientists decided to try using sonar, which measures the echoes from sound waves bounced off objects underwater. With sonar, scientists can determine how large an object is by counting the number of waves they receive.

In 1969, a six-person submarine called *Pisces* was equipped with sonar so that it could search for the monster at depths far below where normal fish live. During one of its cruises, the vessel's sonar detected a large, moving target.

R. W. Eastaugh, the captain of the *Pisces*, said, "A sonar target was picked up whilst *Pisces* was hovering fifty feet off the bottom. *Pisces* homed on the target and when at a distance of four hundred feet the target rapidly disappeared from the screen."

Was it Nessie swimming swiftly away from the submarine? Over the coming years, sonar searches of the loch would make forty contacts with objects that appeared to be large creatures.

In 1987, a team of scientists launched a high-tech search for Nessie called Operation Deepscan. The project used a team of twenty-four boats, all equipped with sonar scanners. As the boats slowly moved together, each lined up

side by side, they formed a giant sonar net that covered the entire width of the loch. On the first day of the mission, the sonar on the boats detected three large creatures swimming below them. They were smaller than whales but larger than sharks. Unfortunately, the ink-black waters of Loch Ness prevented the scientists from being able to tell exactly what the creatures looked like.

The Lost Loch Ness Monster

Before the *Pisces* submarine used sonar to search for Nessie, it served a very different purpose. The sub was originally brought to Loch Ness in 1969 to tow a thirty-foot model of the Loch Ness Monster that was created for a feature film called *The Private Life of Sherlock Holmes*. The *Pisces* pulled the five-ton fake Nessie around the loch, but an accident caused the model to sink

to the bottom of the loch, where it remained undisturbed for almost fifty years.

After the monster model sank, the *Pisces* was outfitted with sonar and began its search for the real Nessie.

While the sonar expeditions did not provide any definite answers, some people wondered if the science of DNA might be a way to solve the mystery of the Loch Ness Monster. DNA is a chemical that is found in the cells of every living thing—from plants to fish to humans—on Earth. If a DNA sample could be taken from Nessie, it could be used to identify her.

In the summer of 2019, a team of scientists from New Zealand made headlines when they announced the results of their DNA study of Loch Ness. After taking over 250 water samples in various locations and different depths of the loch, the team studied DNA floating in the water. This had all been shed by the loch's plant and animal inhabitants. Neil Gemmell, a professor of biology and ecology at the University of Otago in New Zealand, analyzed the data and said that it revealed the presence of about three thousand species. That included fish and other aquatic

Neil Gemmell

species living in the water, such as frogs and toads. It also found thousands of bacteria and DNA from some land-based animals that occasionally swam in the loch, including dogs and humans.

The study made one surprising discovery: It found an unusually large amount of DNA from eels in Loch Ness. Everywhere the team tested, it found eel DNA. Professor Gemmell said that he was not sure if the loch was home to many more eels than anybody had ever estimated or if perhaps giant eels were swimming in the water. No giant eel has ever been caught in Loch Ness, but it's possible that a massive eel could be mistaken for a sea monster.

A puzzling piece of data also emerged from the New Zealand DNA study: The team of researchers did not detect the presence of any seals in Loch Ness. However, seals have been routinely discovered in the loch. It's believed that they swim up the River Ness from the ocean and live in Loch Ness for a few months each year. If the DNA study completely missed the presence of seals in the lake, could it have also failed to detect Nessie's DNA?

Neil Gemmell studies the water of Loch Ness

European Conger Eel

The European conger eel is the world's largest eel and has been discovered to be almost ten feet long, growing to a maximum weight of 350 pounds! Is it possible that some of the many Nessie sightings were of conger eels? Maybe. But one argument against that theory is that eels tend to stay at the bottom of their habitats, where they hide in search of food. With so many Nessie sightings at the surface of the lake, the conger eel is not likely to be the best match for the monster that people have been seeing for years in Loch Ness.

CHAPTER 8
Is Nessie a Plesiosaur?

If Nessie isn't a seal or an eel, then what is she? Some people believe that a huge prehistoric animal found its way to Loch Ness a very long time ago. Could Nessie be some kind of dinosaur?

For those who think that the loch contains a hidden monster, one theory is that a relative of the dinosaur known as a plesiosaur lives in Loch Ness. The plesiosaur was a giant, water-based reptile that lived in oceans more than sixty-five million years ago. Scientists estimate that it may have been forty feet long and weighed almost fifteen tons. It had sharp teeth and ate fish. Some scientists believe that it could also crawl on land. The plesiosaur matches many of the eyewitness descriptions of Nessie, including its long, thin

neck; small head; bulky body with a humped back; four flippers; and long, powerful tail.

For the Loch Ness plesiosaur theory to work, though, Nessie would have had to survive the extinction event that killed off the dinosaurs many millions of years ago. There are several theories about what caused the extinction of the

dinosaurs, including ash from erupting volcanoes or the impact of an asteroid crashing into Earth. Dinosaur fossils, which are the remains of long-dead dinosaurs, have been found on every continent on Earth. From those fossils, scientists have been able to determine how long ago dinosaurs died out.

Plesiosaurs in Print

The idea that dinosaurs and plesiosaurs are still living on Earth received a big boost from two very popular science-fiction novels. One was Jules Verne's 1864 classic *Journey to the Center of the Earth*, in which a team of explorers encountered still-living prehistoric creatures.

The second book was *The Lost World*, a 1912 novel by Arthur Conan Doyle, who created the fictional character Sherlock Holmes. In one part of *The Lost World*, Doyle described an encounter that sounds like many of the Nessie eyewitness sightings:

Here and there high serpent heads projected out of the water, cutting swiftly through it with a little collar of foam in front, and a long swirling wake behind, rising and falling in graceful, swan-like undulations as they went. . . .

"Plesiosaurus! A fresh-water plesiosaurus!" cried Summerlee. "That I should have lived to see such a sight!"

Extinct prehistoric creatures don't usually pop up unexpectedly, but it has been known to happen. The giant prehistoric coelacanth fish was thought to be extinct until one was caught by a fisherman in 1938. If the coelacanth survived, seemingly in secret, why not the plesiosaur?

There are a few problems with the living plesiosaur theory, though. The water in Loch Ness is 42 degrees Fahrenheit, which is way too cold for a reptile like the plesiosaur to exist in. A reptile's body temperature would drop to the same temperature as the loch, and the creature would freeze to death.

Also, Loch Ness was formed by the melting of a giant glacier around ten thousand years ago. That means that a family of plesiosaurs would have had to survive for about sixty-five million years in the North Sea—leaving no trace or fossil remains—until they traveled via the River Ness to start their new home in the loch.

Another weakness in the plesiosaur theory is that in order to continue to breed, scientists estimate that there would have to be ten or more plesiosaurs swimming in Loch Ness. Since plesiosaurs need to breathe air, they would be forced to poke their heads out of the water many times a day. If there are ten or more air-breathing plesiosaurs in the loch, there should have been a *lot* more sightings of Nessie over the years.

The Coelacanth

In 1938, a fishing crew off the east coast of South Africa made headlines when they captured a living fish known as the coelacanth. Marjorie Courtenay-Latimer, a South African museum official, identified and preserved the fish for further study. What made it so remarkable was that the coelacanth swam in our oceans during the Jurassic period, around 150 million years ago! Before 1938, the coelacanth was only known from its fossils and was believed to have died out with the dinosaurs.

The coelacanth is a huge fish, growing to more than six and a half feet long and weighing around two hundred pounds. Its four fins make arm- and leg-like movements, and some scientists believe it could be related to the first creatures that walked on land.

In the words of Oliver Crimmen, a curator at London's Natural History Museum, rediscovering the coelacanth was "the equivalent to finding a dinosaur alive."

Those who are pro–plesiosaur theory have answers to these arguments. They say that the Loch Ness Monster may have evolved over the years, just as many other creatures have done, and it has adapted to life in the loch. Thanks to evolution (the process by which living organisms develop and change over time), Nessie may have acquired new breathing abilities and developed a layer of fat that would allow her to live in the loch's chilly waters. Many also believe that there are dark caverns deep within Loch Ness that have not yet been explored.

Is it possible that Nessie is a cave dweller that only occasionally comes to the surface of the loch?

CHAPTER 9
If Not Nessie . . .

If a massive, plesiosaur-like animal is not swimming around Loch Ness, what other type of creature could it be?

Some argue that the objects in many photos of Nessie are actually water birds, otters, harbor seals, or floating logs. Because Loch Ness is connected to the North Sea by the River Ness, it would be very easy for medium-size sea animals to swim up the river and into the loch.

It's possible that there are large fish living in the loch that could be mistaken for monsters by

people who aren't sure what they're looking at. One potential candidate is the wels catfish, a very big fish that can weigh over six hundred pounds and grow to a length of fifteen feet. A massive fish like the wels catfish living in Loch Ness could explain shadowy shapes seen from above, large disturbances at the surface of the water, and some of the sonar readings from down below.

Some eyewitness sightings of the monster could actually be rippling waves that were caused by wind blowing across the water's surface. Large waves are also created by the many sailboats and motorboats that move through the loch. Could some of the monsterlike humps identified as Nessie's back actually be waves left behind by a passing boat?

It's easy to see how some Nessie sightings could be simple mistakes. People may actually be viewing waves, animals, or other floating objects in the loch. It becomes more difficult to dismiss eyewitness reports, though, when several people at different locations around Loch Ness all report seeing a monster-size creature at the same time.

Sadly, there are also cases of pure hoaxes— tricks intended to fool people. Marmaduke Wetherell's 1933 footprint scam remains one of the most well-known, but unfortunately there have been other hoaxes and practical jokes. Some pranksters have placed floating monster heads and tails in the loch, hoping to fool onlookers. Others have altered photos to make them look like the creature.

In 1993, startling new facts emerged about even the most famous picture of Nessie: the surgeon's photo that had first been published in 1934. Christian Spurling was the stepson of Marmaduke Wetherell, the man who had created the hippo-footprint hoax sixty years earlier. Spurling claimed that he had helped his stepfather fake the surgeon's photo. He said that he had mounted a model of a dinosaur's head and neck on top of a two-foot toy submarine.

Wetherell then took the photo and convinced the well-respected doctor, Robert Kenneth Wilson, to take credit for it. Christian Spurling died shortly after making this confession, and not everyone believed his story.

For those who don't believe the Spurling story, what is known as the surgeon's photo remains the best visual proof that the Loch Ness Monster could be real.

CHAPTER 10
The Loch Ness Monster Today

While the debate continues to rage about the monster's existence, one thing is clear: Nessie is *very* popular.

In 1969, *Encyclopaedia Britannica* revealed that "Loch Ness Monster" was among the leading ten topics on which they received requests for more information. In 1996, Gary Campbell formed the Loch Ness Monster Fan Club after he saw something "unexplained" on the loch. Campbell was disappointed that there was no official group keeping track of Nessie sightings, and so he started the fan club "to record all the sightings and to act as a single point of contact for everybody who's interested in the Loch Ness Monster and her cousins that are in lakes all over the world."

And there is also the Official Loch Ness Monster Sightings Register. This group provides a twenty-four-hour live internet "Nessie on the Net" webcam where viewers can keep an eye on the dark waters of Loch Ness near Urquhart Castle.

There have been Loch Ness Monster–themed cereals, kilts, stuffed animals, food items, and even floor polish. Movies featuring the creature have included *Scooby-Doo! and the Loch Ness Monster, Incident at Loch Ness, The Ballad of Nessie*, and *The Loch Ness Horror*.

Nessie has generated a lot of publicity and a very large amount of money for the country of Scotland. Over one million tourists visit Loch Ness every year, and they spend over $40 million on hotel rooms, Nessie museums, boat rides, and souvenirs.

Nessie isn't the only sea monster in the world. She isn't even Scotland's only giant lake creature. Loch Morar is located southwest of Loch Ness. Loch Morar is a smaller lake than Loch Ness, but it's big enough to host its own monster, called Morag.

Morag

Loch Morar, near Scotland's west coast, is the deepest lake in the country. It is also allegedly the home of a monster known as Morag. Like Nessie, Morag is usually described as a beast with a small head, long neck, and a body that is around forty feet long.

The most famous Morag sighting occurred in 1969 when two men, Duncan McDonnell and William Simpson, were fishing on the loch and saw a dirty-brown-colored "creature" approaching their boat. The beast caught up with them and grabbed the side of the boat in its mouth. McDonnell tried to fight off the monster by hitting it with a wooden oar, but the oar broke. He then grabbed his rifle and fired at the creature. It slowly sank into the water and did not resurface.

In North America, there are legends of other water beasts. Lake Champlain in Vermont has long been rumored to be home to a creature known as Champ. In Lake Erie, which is located between the United States and Canada, there have been sightings of a creature that goes by the name of South Bay Bessie. Within Canada's Okanagan Lake, some say there is a creature known as Ogopogo. Long ago, the Syilx Okanagan people

A possible Ogopogo sighting in Canada

called this water beast N'ha-a-itk, which means "water demon." It was said that N'ha-a-itk had supernatural powers, including the ability to create deadly storms and whirlpools when it was angry.

It seems that in almost every large body of water around the world, there is a version of something like a Nessie. Adrian Shine, a researcher who has devoted over forty years of his life to studying Nessie, says, "We all want monsters . . . that are bigger than we are, frightening and hidden. But to be hidden, they have to be in lost worlds. And to some extent, Loch Ness is a lost world because of its darkness and its depth."

Shine adds, "If monsters exist, then science . . . has ignored the most exciting wildlife mystery in the British Isles. If there are none, then over a thousand people—including doctors, clergymen, MPs [members of the British Parliament], civil dignitaries, not to mention a saint—may have lied; unthinkable."

Nessie is one of the most studied and most mysterious creatures on our planet. The legend of the Loch Ness Monster has outlived many of its hunters.

And so today the search for the Loch Ness Monster still continues. Centuries after Saint Columba confronted a large creature splashing in the loch, Nessie swims on in the imaginations of millions.

Timeline of the Loch Ness Monster

565 — Saint Columba confronts a giant water beast in the River Ness

1879 — Schoolchildren see a giant gray animal walking on land and into Loch Ness

1919 — Margaret Cameron and her siblings encounter a water beast next to the loch

1933 — Aldie Mackay spots a whale-size beast moving on the surface of Loch Ness

— The *Inverness Courier* prints the first newspaper story about a "monster" in Loch Ness

— Hugh Gray's possible photograph of Nessie appears in Glasgow's *Daily Record and Mail* newspaper

1934 — Dr. Robert Kenneth Wilson photographs a dinosaur-like creature rising out of the loch

1960 — Tim Dinsdale films a creature swimming in Loch Ness

1961 — The Bureau for Investigating the Loch Ness Phenomena is formed

1969 — The *Pisces* submarine uses sonar to look for Nessie

1972 — Bob Rines uses a motorized underwater camera in the loch and photographs what appears to be a creature with a giant flipper

2019 — Scientists from New Zealand conduct a DNA study of Loch Ness

Timeline of the World

541	The first major outbreak of a bubonic plague pandemic begins in the Byzantine Empire
1877	Thomas Edison invents the phonograph, a type of record player
1912	The RMS *Titanic* sinks in the Atlantic Ocean on April 15
1931	"The Star-Spangled Banner" is chosen as the United States' national anthem
1933	Construction begins on the Golden Gate Bridge in San Francisco
	Adolf Hitler becomes chancellor of Germany
	Franklin D. Roosevelt becomes president of the United States
1936	The first modern trampoline is built
1960	The first episode of *The Flintstones* airs on TV
1961	Construction begins on the Berlin Wall
1969	Humans walk on the moon for the first time
1971	The environmental protection group Greenpeace is founded
1993	The first *Jurassic Park* movie is released
2019	The Event Horizon Telescope takes the first-ever photo of a black hole in outer space

Bibliography

*Books for young readers

Campbell, Steuart. *The Loch Ness Monster: The Evidence*.
Amherst, NY: Prometheus Books, 1997.

*Claybourne, Anna. *Don't Read This Book Before Bed*.
Washington, DC: National Geographic, 2017.

*Emmer, Rick. *Loch Ness Monster: Fact or Fiction?* Creature
Scene Investigations. New York: Chelsea House, 2010.

*Flitcroft, Jean. *The Loch Ness Monster*. The Cryptid Files.
Minneapolis, MN: Darby Creek, 2014.

Harrison, Paul. *The Encyclopaedia of the Loch Ness Monster*.
London: Robert Hale, 1999.

*Hile, Lori. *The Loch Ness Monster*. Solving Mysteries with Science.
Chicago: Heinemann Raintree, 2013.

Loxton, Daniel, and Donald R. Prothero. *Abominable Science!:
Origins of the Yeti, Nessie, and Other Famous Cryptids*.
New York: Columbia University Press, 2013.

*Peabody, Erin. *The Loch Ness Monster*. Behind the Legend.
New York: Little Bee Books, 2017.

*Townsend, John. *Bigfoot and Other Mysterious Creatures*.
New York: Crabtree Publishing, 2009.

*Yorke, Malcolm. *Beastly Tales: Yeti, Bigfoot, and the Loch Ness
Monster*. New York: DK Publishing, 1998.